I0074564

ARE INDIVIDUALLY ACQUIRED
CHARACTERS INHERITED?
(1893)
BY
ALFRED RUSSEL WALLACE

Copyright © 2013 Read Books Ltd.
This book is copyright and may not be
reproduced or copied in any way without
the express permission of the publisher in writing

British Library Cataloguing-in-Publication Data
A catalogue record for this book is available from the
British Library

Alfred Russel Wallace

Alfred Russel Wallace was born on 8th January 1823 in the village of Llanbadoc, in Monmouthshire, Wales.

At the age of five, Wallace's family moved to Hertford where he later enrolled at Hertford Grammar School. He was educated there until financial difficulties forced his family to withdraw him in 1836. He then boarded with his older brother John before becoming an apprentice to his eldest brother, William, a surveyor. He worked for William for six years until the business declined due to difficult economic conditions.

After a brief period of unemployment, he was hired as a master at the Collegiate School in Leicester to teach drawing, map-making, and surveying. During this time he met the entomologist Henry Bates who inspired Wallace to begin collecting insects. He and bates continued exchanging letters after Wallace left teaching to pursue his surveying career. They corresponded on prominent works of the time such as Charles Darwin's *The Voyage of the Beagle* (1839) and Robert Chamber's *Vestiges of the Natural History of Creation* (1844).

Wallace was inspired by the travelling naturalists of the day and decided to begin his exploration career collecting specimens in the Amazon rainforest. He explored the Rio Negra for four years, making notes on the peoples and languages he encountered as well as the geography, flora, and fauna. On his return voyage his ship, Helen, caught fire and he and the crew were stranded for ten days before being picked up by the Jordeson, a brig travelling from Cuba to London. All of his specimens aboard Helen had been lost.

After a brief stay in England he embarked on a journey to the Malay Archipelago (now Singapore, Malaysia, and Indonesia). During this eight year period he collected more than 126,000 specimens, several thousand of which represented new species to science. While travelling, Wallace refined his thoughts about evolution and in 1858 he outlined his theory of natural selection in an article he sent to Charles Darwin. This was published in the same year along with Darwin's own theory. Wallace eventually published an account of his travels *The Malay Archipelago* in 1869, and it became one of the most popular books of scientific exploration in the 19th century.

Upon his return to England, in 1862, Wallace became a staunch defender of Darwin's landmark work *On the Origin of Species* (1859). He wrote responses to those critical of the theory of natural selection, including 'Remarks on the Rev. S. Haughton's Paper on the Bee's Cell, And on the Origin of Species' (1863) and 'Creation by Law' (1867). The former of these was particularly pleasing to Darwin. Wallace also published important papers such as 'The Origin of Human Races and the Antiquity of Man Deduced from the Theory of 'Natural Selection" (1864) and books, including the much cited *Darwinism* (1889).

Wallace made a huge contribution to the natural sciences and he will continue to be remembered as one of the key figures in the development of evolutionary theory.

Wallace died on 7th November 1913 at the age of 90. He is buried in a small cemetery at Broadstone, Dorset, England.

ARE INDIVIDUALLY ACQUIRED CHARACTERS INHERITED? (1893)

I.

The question which forms the title of this article will not, perhaps, strike the general reader who is unacquainted with recent developments of biology as being of much importance, or as having any special interest for the world at large. Yet it really involves considerations hardly less far-reaching than evolution itself, since the correct answer to it must depend upon, and be, a logical consequence of a true theory of heredity. If, then, we can arrive at this correct answer, either by means of observation of natural phenomena or by experiments with living organisms, we shall possess a criterion by which to judge between rival theories; while the answer itself will be found to have a direct bearing of a very important kind on possibilities and methods of human improvement.[1]

Up to about ten years ago the answer to the question would have been almost unanimously in the affirmative. Darwin accepted the inheritance of such characters as an undoubted fact, though he did not attach much importance to it as an agent in evolution; and his theory of pangenesis was an attempt to explain the phenomena of heredity in accordance with it. Mr. Francis Galton made some experiments which led him to doubt the correctness of Darwin's main contention--that minute gemmules from every cell in the animal body were collected in each of the germ and sperm cells, and thus led to the reproduction of a similar body. He transfused large quantities of blood from black to white rabbits, and *vice versâ*, without in any way injuring them; and after this infusion of blood from a very distinct variety, sometimes to the extent of one-third of its whole amount, each kind bred as true as before, showing no signs whatever of intermixture. He then developed a new theory of heredity, which appeared to him more in accordance with facts, and an essential part of this theory was that the germinal material passes direct from parent to offspring, instead of being produced afresh from the various parts of the body; and, as a consequence, changes produced in the body by external agencies during its life will not be transmitted to the offspring. A few years later Professor Weismann, of Freiburg-in-Baden, independently arrived at a somewhat similar theory, founded on the embryological researches of himself and other biologists; and he supported it by such a body of evidence and by such a wealth of illustration and reasoning that it at once attracted the attention of biologists in every part of the world. This theory being manifestly opposed to the inheritance of acquired characters, he was led to examine the evidence for this dogma, and found it to be extremely scanty, and for the most part quite inconclusive. But as some biologists of great eminence believe that the inheritance of such characters is absolutely necessary in order to explain many of the phenomena of evolution, the discussion on this point has been carried on by many who would have felt little interest in the problem if it were one of embryological development alone. Year by year the question has been discussed in books, pamphlets, and review articles, while Professor Weismann has continued his studies on the whole subject, and in a volume of which an English translation has just appeared, has worked out his theory in very minute detail, grappling fairly with all the various phenomena to be explained, and thus putting the whole question before the scientific world in a manner which allows it to be fully discussed, tested, and controverted.[2]

This detailed theory is far too complex and technical to be explained in a review article; but as its truth implies that the inheritance of acquired variations is not a law of nature, and as Mr. Herbert Spencer has just set forth some fresh arguments in favour of such inheritance, and has also reinforced some of his former arguments (in two articles in the *Contemporary Review*), while an American naturalist has just issued a work,[3] in the introduction of which he discusses the same question, and summarises what seem to him the strongest arguments that have been advanced on both sides, concluding also in favour of the inheritance of such characters--a good opportunity is offered to review this evidence, and to show, as the present writer thinks he can show, that all the alleged facts and arguments are inconclusive, and that the balance of the evidence yet adduced is altogether in favour of such characters not being inherited.

It is first necessary to understand clearly what is meant by "acquired characters," as even naturalists occasionally miss the essential point, and take any peculiarity that appears in an individual during life to be an "acquired character." But such peculiarities are usually inherited from some ancestor, unless they can be clearly traced to special conditions to which the individual's body has been exposed. As an illustration, let us suppose twin brothers, very similar in all physical and mental characters, to be subject during life to very different influences: one being brought up from childhood to city life and kept closely at a desk till middle age, the other living always in the country and becoming a working farmer. If the one were then pale, slender, weak, and delicate, the other ruddy, stout, and strong, these differences would be, in one or the other, probably partly in both, "acquired characters." And if both, at the same age, married twin sisters who had been each subject to corresponding conditions, the common idea is that the children of the city couple would be inherently weakly, those of the country couple strong; and that the balance would not be restored even if these two families of children were subject, during their whole lives, to identical conditions. In other words, it is usually believed that the acquired characters of the parents would be transmitted to the constitutions of the children. But it is now asserted, by Weismann and his followers, that facts do not agree with this assumption, and that, in the case supposed, both sets of children would inherit the original qualities of the parents, modified, perhaps, by qualities or characteristics of remoter ancestors, but not showing any effects of the changes produced in their parents by external conditions only.

This latter belief is, I am informed, held and acted upon by breeders of animals as the result of their extensive experience. If a young dog or horse of high breed and good external points becomes accidentally lamed, so as to be permanently disabled from the usual work of its kind, it is often kept for many years to breed from, in full confidence that its offspring will inherit the good qualities of the stock, and will be in no way deteriorated by the absence of work calculated to strengthen the muscles, enlarge the chest, and otherwise increase the power and activity of the parent.

Again, if the effects of the use of certain muscles, or of special mental faculties with their corresponding nervous and muscular co-ordinations, were transmitted to offspring, then definite results ought to have been so frequently produced as to have become embodied in general experience and popular sayings. Take the case of any mechanic working at his trade, whether blacksmith, carpenter, watchmaker, or any other art leading to the use or disuse of special muscles or faculties. If long-continued exercise in one direction leads to increased strength and skill which is passed on to the children, then it ought to be an observable fact that the younger sons should have more strength and skill in their father's business than the firstborn; but, so far as I know, this has never been alleged. So with men of genius, whose mental faculties have been fully exercised in special directions, whether as men of science, artists, musicians, poets, or statesmen; if not only the inherent faculty but also the increased power derived from its exercise be inherited, then we ought frequently to see these faculties continuously increasing during a series of generations, culminating in some star of the first magnitude. But the very reverse of this is notoriously the case. Men of exceptional genius or mental power or mechanical skill appear suddenly, rising far above their immediate ancestors; and they are usually followed by successors who, though, sometimes great, rarely equal their parent, whose pre-eminent powers seem generation after generation to dwindle away to obscurity. As illustrations of this principle we may refer to such men as Brindley, Telford, Stephenson, Bramah, Smeaton, Harrison (inventor of the chronometer), Brunel, Dollond, Faraday, Wren, John Hunter, and many others, who were mostly self-taught, and derived nothing apparently either from the faculties or the acquired powers of their parents. So almost all the great poets, musicians, and artists of the world start up suddenly and leave no equals, far less superiors, among their offspring or their descendants. These various classes of facts not only lend no support to the theory of the transmission of acquired faculties from generation to generation, but are not what we should expect if such transmission were a fact. They certainly serve to throw doubt upon it and to show that inheritance is not such a simple matter as this theory implies; they may, therefore, prepare the reader to consider with impartiality the facts and arguments that have been put forward in its favour, together with the reasons I shall give for the inadequacy of those arguments. For it must always be remembered that the inheritance of this class of characters or qualities must be proved by facts that admit of no other interpretation, if it is to be accepted as one of the bases of the theory of organic evolution. When such tremendous issues are at stake we cannot base our faith upon probabilities, unless they are to an overwhelming degree greater than the probabilities on the other side.

I propose to waste no time on the question whether mutilations are ever inherited, because both parties are now agreed that this is not the point at issue. What we want to know is, whether the effects produced during the lives of individuals by such natural causes as the use or disuse of certain muscles or organs, change of food, or change of climate, are transmitted to offspring, so as to accumulate such effects and thus serve as an important factor in evolution.

Two of the cases which have been adduced as affording very strong evidence of the inheritance of an acquired character are, the habit of dogs to turn round several times before lying down, and the peculiar play-habits of the bower-birds; these being supposed to be beyond the power of natural selection to produce, because neither are of vital importance to the species. But such cases as these really prove nothing, because so much in them is hypothetical. It is only guessed that the dog's habit is derived from wild dogs turning round to make a comfortable bed in rough grass. But even if this be a fact, there are many awned seeds of grasses which prick the skin, and in some cases work their way into the body, causing wounds or death, and the turning round may have the effect of laying these awned seeds parallel to the hairs and thus prevent them from penetrating the skin. If so, natural selection would produce and preserve the habit. Again, it may, with many dogs, be a matter of simple comfort, the turning serving to keep the rather stiff hair all the right way, and also to brush away small hard objects. Yet again, it is not alleged that all dogs do it, and in many cases it may be a habit copied from the mother. The uncertainties of the case are therefore too great for it to afford an argument of any value. The bower-birds' habits are more difficult to explain on any theory, since the whole question of these alleged instincts is unsettled. We have evidence that in many cases even the peculiar song of birds is not instinctive in the species but is the result of imitation; and Mr. Hudson has recently shown that the fear of man in wild birds, or its absence, is probably the result of individual experience in all cases. Till we know that the bower-birds' habit is wholly due to inheritance and not to imitation of older birds, we can hardly found any important conclusions upon it.

Many writers have laid stress on the difficulty of accounting for the origination of new organs in certain groups of animals, by variation and natural selection alone. Horns are especially adduced; and it is alleged that there is no other way of explaining their origin except by the habit of butting with the head, leading to thickening of the skin and excrescences of the bone, which, being transmitted by inheritance and increased by use, gradually produce the various kinds of horns. In like manner, the origin of flowers and their successive modification, have been imputed to the irritation caused by insects, leading to outgrowths which have been inherited and increased by further irritation.

Taking the case of the horns, Mr. J. T. Cunningham, in his introduction to the English translation of Eimer's *Organic Evolution*, says:--"No other mammals have ever been stated to possess two little symmetrical excrescences on their frontal bones as an occasional variation; what then caused such excrescences to appear in the ancestors of horned ruminants? Butting with the forehead would produce them, and no other cause can be suggested which would."

This assertion, that "butting with the forehead would produce them," assumes the whole question at issue. There is, I believe, no evidence of it whatever, and there is much that renders it improbable. And the first part of the statement is also erroneous, for Darwin tells us, "In various countries horn-like projections have been observed on the frontal bones of the horse; in one case described by Mr. Percival, they arose about two inches above the orbital processes, and were very like those of a calf from five to six months old, being from half to three quarters of an inch in length." As no known animal in the ancestral line of the horse had horns, these must have been "new characters;" and had they appeared before the ancestral horses acquired such powerful weapons of offence in their hoofs, they would probably have been preserved and increased by selection into formidable weapons. That horns have not unfrequently arisen from such apparently uncaused variations, is indicated by the remarkable difference of structure and growth in the horns of such nearly allied groups as the deer and the antelopes, which at a quite recent epoch must have originated independently. Very suggestive is the curious enlargement of the skull under the crest of the Polish fowls. In another fowl's skull, figured by Darwin, there are two large rounded knobs on the forehead, forming perfect incipient horns.

Dermal appendages, which could not have been caused

by special irritations, are so frequent, that almost any useful development appears possible. The spines of the hedgehog and the quills of the porcupine, are of this nature, as are the plates of the armadillo and the scales of the pangolin. The feathers of birds are one of the most marvellous of these developments which, when they once arose, were preserved and modified in endless ways. So, the curious erectile appendage on the forehead of the South American bell-birds, and the equally strange feather-covered cylinder pendent from the throat of the umbrella-birds, are other illustrations of these abnormal outgrowths of the skin for the origin of which we can assign no cause. Many other secondary sexual appendages of birds are equally inexplicable in their beginnings, such as the long feathers springing from the sides of the head in the six-plumed bird of paradise, and the singular pair of long white feathers growing from among the upper primary wing-coverts of Semioptera wallacei, to which I believe there is nothing similar in the whole class of birds. These various cases of dermal appendages are sufficient to indicate that variations of this kind are continually occurring, which, whenever useful, have been seized upon and increased by natural selection, since any such variations appearing among our domesticated animals are found to be strictly inherited.

The American naturalists lay much stress on the evolution of the teeth of mammals in complete palæontological series, alleging that the successive modifications of the cusps conform strictly to lines of use and disuse, and that they are therefore produced by use and disuse. To this there are two distinct replies, either of which seems to me sufficient. In the first place, in such vitally important organs as the teeth of mammals, natural selection will necessarily keep them on these lines, because *use* implies *utility*, and *disuse, inutility*, and utilities necessarily survive. If, then, variations occur in the forms of the cusps--and they certainly do--natural selection will modify them along these lines of utility; and it will be absolutely impossible, from a study of the series of fossil forms, to prove that they have been directly modified by use and that the modifications have been inherited, and that they are not the result of normal variations accumulated by survival of the fittest. The second reply is that of Mr. Poulton, who points out that the form of the tooth is fixed before it cuts the gum and that use only wears the cusps away. It is therefore difficult to see how such use in the parent can determine any definite variation in the teeth of the next generation.

There is also a general argument, in the fact of so many special types of teeth having been developed, which cannot have been produced by the corresponding use. Such are the arrangement of enamel and dentine in the incisor teeth of rodents, so that they preserve a continual chisel-like cutting edge, and, unlike the teeth of most other mammals, keep on growing at the root so that they are pushed up as fast as they wear away; and the remarkable molar teeth of the elephants, which come forward in succession, and by the arrangement of the folds of hard and soft material always keep a grinding surface, while the enormous tusks grow on continuously during life. These and many other singular modifications of teeth can certainly not be traced to corresponding diversities of use as directly producing them, while they are easily explained by the great variability of all complex organic structures, furnishing material for endless modifications according to the various needs of the widely different mammalian types.

We have now to consider Mr. Herbert Spencer's objections in the articles already referred to, which bear the formidable title, "The Inadequacy of Natural Selection." The first of these objections is founded on Weber's experiments on the sense of touch, showing that the power of distinguishing the sensations produced by two points rather close together varies greatly in different parts of the body, the tips of the fingers being able to distinguish the points of compasses when the twelfth of an inch apart, while on the middle of the back they have to be opened more than two inches in order that the pressure of two points may be distinguished from that of one. Between these extremes tactual discrimination varies in different parts of the body, apparently without much relation to utility, except of course in the case of the fingers;

and after detailing these at some length Mr. Spencer asks how these divergences can possibly be explained by natural selection. "Why," he asks, "should the thigh near the knee be twice as perceptive as the middle of the thigh?" And he urges that, in order to prove that these many small variations in different parts of the body have been produced by natural selection, it must be shown that they have influenced self-preservation. He then goes on to show that this perceptive power can be increased by exercise; and concludes with a theory that the differences of tactual perception in different parts are proportioned to the amounts and varieties of contact with substances to which they are subjected, and that these variations in amount and kind of contact have produced diversities of sensibility in the individual, which, by inheritance, have been accumulated in the offspring.

Now, this whole inquiry, and the conclusions drawn from it, seem to me (with all respect for Mr. Spencer's great abilities) to afford a glaring example of taking the unessential in place of the essential, and drawing conclusions from a partial and altogether insufficient survey of the phenomena. For this "tactual discriminativeness," which is alone dealt with by Mr. Spencer, forms the least important, and probably only an incidental portion of the great vital phenomenon of skin-sensitiveness, which is at once the watchman and the shield of the organism against imminent external dangers. The sensations we receive by means of the skin, of contact with various substances, smooth or rough, blunt or pointed, dry or wet, cold or hot, whether indifferent, or pleasurable, or painful, or agonising, afford us information and safeguards without which we could, in a state of nature, hardly exist uninjured for a single day. And we shall find that the delicacy of this absolutely essential danger-signal is almost exactly in proportion to the vital importance of the part to be protected from danger. Thus the maximum of sensitiveness is found in the eye and its surrounding membranes, not because they are most frequently subject to a variety of contacts, for the very reverse is the case, but because this organ is at once the most delicate and the most important for the safety of the individual. So the hands and feet are not very sensitive in those parts which are specially adapted to come in contact with external objects, but in those parts where the tendons, nerves, and bloodvessels which render them effective organs are most exposed to injury, as in the palm of the hand and the hollow of the foot, and especially on the under side of the arm just above the wrist. The parts of the body which are less sensitive are those where there are masses of muscle, the puncture of which would not cause any serious injury. It will thus be seen that sensitiveness to pain from external agencies is not at all proportioned to the frequency of contacts, but to the vital importance of the parts to be protected; and it is, therefore, such as could not possibly have been produced by inherited use, but must have been developed solely by the preservation of useful variations; and as it is essentially a life-preserving faculty this would inevitably have been effected.

It seems most probable that the faculty of tactual discrimination of adjacent points is partly an incidental result of the distribution of nerve-endings required by skin-sensation as a preservative faculty, and partly the result of use and attention in the individual. All the facts adduced by Mr. Spencer are in accordance with this view, while none of them in the slightest degree necessitate inheritance of individual experiences to account for them. To show any probability of such inheritance it must be proved that this tactual discriminativeness is a special faculty, due to a different set of nerves from the known nerves of sensation, which, I believe, has never even been suggested. But if it is due to the same nerves, then to separate this small and comparatively unimportant function of skin-sensation from the great and vitally important functions the faculty subserves, and to found upon this artificially isolated and unimportant fragment an argument against the adequacy of natural selection, is not only quite inconclusive, but, as an argument, is altogether null and void.

II.

The next point discussed by Mr. Spencer is the mode in which the eyes of the Proteus of the Carniola caverns have been reduced to a rudimentary condition; his argument being, that, unless the reduction in size by disuse was inherited, there is no means of accounting for the almost total disappearance of the eyes. It may be stated that the views held by the followers of Weismann in this country are, first, that as natural selection is always at work to keep all important organs when in use up to their full size and efficiency, the withdrawal of natural selection when the organ ceases to be used, termed by Weismann *panmixia*, will, by allowing the most imperfect as well as the most perfect eyes to survive, reduce the average size and quality considerably. Then comes the consideration that in total darkness such a delicate organ as the eye would be subject to frequent injury, producing inflammation, and either directly or as a secondary result, death; thus again weeding out those with the largest and most prominent eyes, or those which kept them open habitually in the effort to see. And, lastly, there would come into action what is called economy of growth, the diminution of any useless, but complex, organ being beneficial, owing to the saving of material and growth-power in building it u[4] It is with this last factor that Mr. Spencer deals, and endeavours to show that the economy would be infinitesimal, because the weight of the eye of this animal is so small. He supposes the reduction to be effected in two thousand years by decrements of one two-hundredth part every ten years; and taking the original eye to have weighed ten grains, he almost laughs to scorn the idea that such an almost infinitesimal amount of diminution at any one time could have given the animals in which it thus diminished a greater chance of survival.

Now, there are two very serious oversights which entirely invalidate this argument. The eye is treated as if it were mere protoplasm weighing so many grains, instead of being a highly complex organ with which muscles, blood-vessels and nerves are connected and co-ordinated in greater proportion perhaps than in any other organ. I presume the original eye of the ancestral Proteus must have had its three distinct sets of nerves--those of vision, of sensation, and of motion--involving in their normal use the expenditure of a considerable amount of nervous energy, besides the various muscles and bloodvessels connected with it. To measure the benefit to be derived from the entire suppression of such a complex organ, when it became useless, as no more than from the gain of so many grains of simple muscular tissue, appears to me to be an extraordinary misconception of the conditions of the problem.

The second oversight is in ignoring the tremendously severe struggle for existence that would necessarily arise when an animal which had heretofore had the full use of eyes in obtaining food, avoiding danger, and finding its mate, had to enter upon a perfectly new kind of existence, in total darkness, and, moreover, in a place where all kinds of vegetable or animal life were so scanty that the wonder is how those individuals who were first carried into the cavern escaped starvation. Under such conditions as these, would not the various modes of reduction of the eyes above suggested act with an energy and rapidity far beyond their action under normal conditions? And might we not expect the most extreme variations in the direction of the abortion of the eye and of its connected tissues, muscles, and nerves, to have an exceptional value when the food required for building up the organism could only be obtained with the greatest difficulty and in the most limited quantities? Under such conditions I should not be surprised if the greater part of the actual eye-reduction had been effected in fifty or a hundred years, instead of in two thousand. But whether this were the case or no, it will, I think, be admitted that to ignore all these very exceptional conditions, and to argue the case as if the whole matter were one of the economy of a few grains of tissue to an animal whose food-supply was normal, does not add anything to the evidence for the inheritance of acquired variations.

The only other argument of importance adduced by Mr. Spencer is that drawn from the giraffe. This argument was first stated at some length in his essay on *The Factors of Organic Evolution*, in 1886; and it is now repeated and enforced by some additional considerations, although, Mr. Spencer says, he has met with nothing that can be called a reply; and he adds, that my contention (in *Darwinism*) that what he alleges cannot be done by natural selection has been done by artificial selection--assumes a parallelism that does not exist. I therefore propose to examine this case more carefully, and shall show that the parallelism I assume is a very close one, and that natural selection is, as Darwin himself believed, fully competent to account for the facts.[5]

Mr. Spencer's argument is, briefly stated, that to develop such an animal as the giraffe from some antelope-like ancestor requires many coincident and co-ordinate variations of different parts--each increase in the length of the neck, of the head, of the fore or hind limbs, or of either of the bones composing them, requiring corresponding increase of muscles, nerves, and blood-vessels, not only of such as are immediately connected with the enlarged limb, but often in remote parts of the body whose motions are necessarily co-ordinated with it. He maintains that any increase of one part without the adjusted increase of other parts would cause evil rather than good; and that want of co-adaptation, even in a single muscle, would cause fatal results when high speed had to be maintained while escaping from an enemy. Then, again, not only the sizes but the shapes of the bones have to be altered as the muscles are increased in size and the motions of the various parts of the body change; and this introduces fresh difficulties which are, again and again, declared to be insurmountable. And after elaborating all these alleged difficulties at great length, he arrives at the conclusion that, unless the increase or modification of parts, due to use by the individual, is inherited, there can have been no evolution.

Now, I believe, and hope to be able to prove, that these accumulated difficulties are almost wholly imaginary, and arise from a neglect to consider known facts of variation and known methods of adaptation. Mr. Spencer accepts the fact that I and others have laid stress upon, that individual variations are continually occurring in all parts of the organism and in all directions, and that the variations of each part are often independent of each other; but he ignores the equally undoubted fact that certain parts are correlated, and very often do vary simultaneously. The diagrams I have given in my *Darwinism* show this clearly. A considerable number of parts, as the wings and tail, tarsi and toes of birds, usually vary together, either to the same or to a different amount; but all of them sometimes vary independently, and even in an opposite direction to each other; and such irregular variations evidently afford the very best material for natural selection to work upon, since any kind of variation, either coincident or independent, can be rapidly accumulated. This fact alone does away with half of Mr. Spencer's difficulties.

Another considerable portion of the supposed difficulties is created by assumptions which pervade his articles but which are opposed to the facts of nature. He tacitly assumes that natural selection works by the preservation of large individual variations--"sports" as they are often termed; whereas both Darwin himself and all his followers entirely reject these as causes of modification of species (except perhaps in rare cases where they may initiate new organs), and rely wholly on those individual variations which occur by thousands and tens of thousands in every generation. Mr. Spencer continually uses such expressions as, "This one has unusual agility"; "that one develops longer hair in winter"; "another has a skin less irritated by flies"; "it is needful that the individual in which it occurs shall have more descendants"; "a variation . . . might sensibly profit the individual in which it occurred"; "would an individual . . . survive"; "favourable variations . . . would disappear again long before one or a few of the co-operative parts could be appropriately varied." The same assumption pervades the writings of most of the opponents of Weismann. Thus Mr. Keeler, in the work already quoted, says, referring to the modification of organs by variation and selection only: "This explanation is open to two objections: first, the one already raised by Mr. Spencer, that before the second correlative variation appeared the first would be lost; and, second, the suggestion of Cunningham in regard to assuming that use could develop the character required, but that the individual thus favoured could not transmit the variation, but that posterity must wait for the same variation to arise spontaneously. This hypothesis is so forced, illogical, and absurd, that so long as a better one can be found it should be adopted."

Now all this implies that there are but few variations occurring at long intervals; the facts being that, in populous species, every generation affords many thousands of variations of every observable part and organ; whence it follows that the individuals of every such species can be divided into two sets as regards each organ or each group of parts, such as those with a longer and those with a shorter neck, those with neck and legs longer and those with the same parts shorter. In the latter case perhaps a quarter or an eighth only of the whole population would be found in each category; but as, in most cases, not one eighth part of those born each year can survive, this would be ample. It will be seen that the facilities for modification are thus indefinitely greater than the expressions and arguments of Mr. Spencer and his supporters assume them to be.

Another tacit assumption is, that in nature all the individuals of a species have their parts so perfectly co-ordinated that any increase of one part only would disturb the harmonious adjustment and be a disadvantage. But this is totally to misconceive the situation. The adjustment of parts is a rough working adjustment, sufficient for the purpose of maintaining life, but capable of being improved (or deteriorated) by very many slight modifications of single parts. To illustrate this general adjustment, let us suppose we have before us for comparison all the county elevens of English cricketers. We shall have a body of some hundreds of picked men, all of whom are probably above the average as runners, are exceptionally quick with eye and hand, and are all more or less active and muscular. They vary, of course, in their special capacities, whether as batters, bowlers, fielders, or wicket-keepers, but it is certain that most of them would take a high place in almost any form of athleticism to which they chose to apply themselves. Yet these men would not resemble each other closely in stature or proportions. We should find among them tall and short, slender and stout; and among those of the same height proportions would differ, some being long-legged, others short-legged, and perhaps no two of the lot would be found to have exactly the same proportions in all measurable parts of the body. We are thus shown that a high average result of strength and activity can be reached by very various combinations of the bones and muscles of the limbs and other variable parts, and we can hardly doubt that almost all of these men could be rendered still more efficient cricketers or athletes by some slight improvement in their organisation. One would run better if his legs were longer, another would throw and bowl better if his arms were shorter and more muscular; and such changes would be effective because these parts are now imperfectly co-ordinated with the rest of the body.

The considerations suggested by this illustrative case immensely increase the facilities for the improvement of any faculty required by natural selection, and they enable us to understand the process by which both natural and artificial selection have been able to modify the form and qualities of so many animals. It is not, as Mr. Spencer's argument assumes, by the selection of improvements in any special bone, or muscle, or limb that these modifications have been effected, but by *the selection of the capacities or qualities resulting from the infinitely varied combination of variations that are always occurring.* Horses have been improved by preserving the swiftest for racing, the strongest for the plough, and a combination of speed, endurance, and jumping for the hunting field. Improved combinations do occur in all these classes in every fresh generation; and breeders know that from, say a hundred half-wild horses, whether from the prairies, the pampas, or the Australian bush, careful selection could obtain--probably in less than a century--very excellent representatives of each of the three above-mentioned types of horses.

Recurring now to the case of the giraffe, whose whole organisation has been modified so as to obtain an almost unlimited feeding-ground from bushes and trees far above

31

the limits eaten off by the tallest antelopes and buffaloes, we see that numerous successive modifications have all worked in this one direction--the large size, the enormous shoulders giving a sloping back, the correspondingly long neck, the long head, and the elongated prehensile tongue, so that a full-grown male giraffe can feed up to twenty feet above the ground.

In considering how this was brought about we must remember that the struggle for existence is very intermittent in character. A severe drought causing a scarcity of herbage, and leading the antelopes and buffaloes to compete with the giraffes for the foliage near the ground, may occur at intervals of from twenty years to a century. When such droughts occur, animals that can feed half a foot higher than others would survive; not one individual here and there, but the tenth or twentieth part of the whole population in a given district, amounting, perhaps, to many thousands of individuals. These taller survivors would go on increasing for another long period, in the meantime being subject to the usual struggle with wild beasts and other dangers, natural selection keeping the whole organism up to the mark, till another period of scarcity led again to an elimination of the shortest. Those that survived would not be all alike; some would gain the increased feeding power by higher shoulders, some by longer neck, or longer head, or more extensible tongue, or by various combinations of these variations; and the regular elimination of the weaker or less active in the intervening periods of abundant food, together with the equalising result of constant intercrossing, would produce a race possessing the one essential character of lofty feeding power by means of the altogether unique combination of characters found in the giraffe. There has been ample time for the process, probably the whole of the pliocene period; and in view of the known amount of variability in all parts of the organism, and the intermittent nature of the struggle for existence, there seems far less difficulty in the production of this animal than in the production of the beautiful and perfectly co-ordinated greyhound in an almost infinitely shorter time.

And now we see how very close is the parallelism between artificial selection and natural selection, and how purely imaginary are the difficulties of co-ordination set forth so elaborately by Mr. Spencer. For, by the preference of some men for swiftness in their dogs, of others for those which followed game by sight rather than by scent, of others again for elegant slenderness of form, the greyhound, almost as perfect as we now see it, had been already developed in the time of the ancient Egyptians from some wolf-like wild ancestor. Men have always preserved qualities, not single characters, just as nature preserves the qualities of speed, strength, agility, acute scent or vision, not any particular variation of bone, muscle, or member. The two modes of selection are thus strictly analogous and strictly comparable; and the whole elaborate structure of "insurmountable difficulty" founded by Mr. Spencer on the supposed impossibility of adjustment of parts by variation and selection, falls to the ground. As a matter of fact, there is a sufficiency of *useful* variation always present in each succeeding generation to increase any required life- preserving quality, all theoretical objections to the contrary notwithstanding.

In his *Principles of Biology*, Mr. Spencer discussed, briefly, the smaller jaws of civilised races as only to be explained by the inheritance of successive reductions produced by the use of softer food, since any advantage to be gained by each step of the process would be too minute to have any appreciable effect in saving life under any possible circumstances. In his *Factors of Organic Evolution* he recurs to the subject, in order to exclude two imaginable causes which he had not before referred to--a correlation of decreased jaw with increased brain, which he rejects, because many small-brained people are also small-jawed, while others distinguished for their mental power have yet jaws above the average size: and sexual selection, which he also rejects for obvious and valid reasons.

Subsequently, the same question was discussed by Mr. W. Platt Ball, in his excellent little work, *Are the Effects of Use and Disuse Inherited?* He shows that the difference of size is less than has been supposed, while the great range of variation in the size and weight of the jaw, both in civilised and savage races, renders all comparisons useless unless founded on a very large number of specimens. He then accounts for the reduction that actually exists by a variety of suggested causes, some of which are undoubtedly open to criticism. He also states that he "allows for a reduction proportional to that shown in the rest of the skull," because some other agency than disuse must have reduced the thickness of the skull, and, presumably, of the jaws also; after which allowance there remains but a small further reduction of the jaws to be accounted for.

This portion of Mr. Platt Ball's work called forth a very minute criticism from Mr. F. Howard Collins, which, being declined by the editor of *Nature*--as the author tells us in his preface--has been published in pamphlet form. As Mr. Collins is the author of *An Epitome of the Synthetic Philosophy*, so well done as to have received the approval of Mr. Spencer himself, he may be supposed to speak with some authority when he undertakes to attack one of Mr. Spencer's critics; and for this reason I think it advisable to point out the erroneous and illogical principles on which his whole argument is founded. In order to compare the jaws of Australians with those of recent Englishmen, he takes one linear dimension of the jaws, cubed to get the proportionate bulk, and compares this with the cubical contents of the skulls to which the jaws belong, that is, with the mass of the brain. Now, as the Australians have very much smaller brains than Europeans, to compare the jaw with the skull-capacity is to make it appear much larger in the Australian and smaller in the European, even when the actual size is identical. Why should the jaw be measured by the brain as a standard, any more than the foot, the hand, or the stature? On the same principle the Australians might be proved to have very large hands, and to be much taller than Englishmen, whereas, like most savages, they have rather small hands, and are hardly our equals in stature. The only reasonable way of comparing the jaws of two races so nearly equal in stature as Australians and Englishmen, is to compare them directly. The jaw is really a limb, used in the mastication of food to supply the whole body with nourishment, not the brain only; and its true organic relation is with the body, not with the brain-case to which it is attached, still less with the brain itself. By the illogical process he uses, of first increasing the linear dimension of the Australian's jaw in the same proportion as the English skull is larger than the Australian skull, without attempting to ascertain whether the depth and thickness increase in the same proportion, and then cubing this dimension, he arrives at the amazing result that the Australian's jaw is very nearly double the bulk of the modern English jaw! Mr. Platt Ball, on the other hand, finds the weight of the latter to be only 5 per cent. less, when reduced in proportion to the lighter skull, the actual difference being 17 per cent.; but both results are founded on far too small a number of specimens to be in any way trustworthy. Mr. Collins then replies, and I think forcibly, to two of Mr. Ball's suggested causes of the decrease--lightness of structure facilitating agility, and sexual selection--but he passes over the two which have most weight, the one as "a tentative suggestion," the other altogether unnoticed; namely, cessation of selection, and increase or decrease caused by use and disuse in the individual. The first, Mr. Collins says, with a strange misapprehension of the point, would only affect the weight and thickness of the jaw, not the peripheral measurement which he has used. Buy why should not the size--the length and depth--of the jaw be quite as important as the thickness and strength? The greater length and depth of the jaw would be effective in giving more room for the attachment of the muscles on which the whole efficiency of the masticating organ depends, as well as by affording space for the full development of large and well-formed teeth. In the early stages of human progress, when much indigestible food had to be eaten, large bones to be gnawed, and meals to be hastily devoured, the large and powerful jaw would be preserved by natural selection. But civilised man has no need for such a bulky apparatus, hence the average size of the jaw would fall "from the birth-mean to the survival-mean," to use Mr. Lloyd Morgan's neat expression--small-jawed men being at no disadvantage in the struggle for existence; whence the occurrence among us of very small and very large jaws, though with a lower average than among the Australians.

The other point--and I believe a very important one--is the diminution, *in the individual*, due to comparative use and disuse. From the time the first teeth are obtained, the jaws are used in mastication many times a day, and the difference in the amount of exertion and strain on the muscles and bone, in the case of a civilised European living mostly on soft or well-cooked food, and the savage chewing up tough roots and tearing half-raw flesh from the bones of almost any animal he can kill, must be very great. This difference acting while the bones of the face are growing, in the period between childhood and manhood, and to a less degree, perhaps, on to middle age, would certainly lead to a difference of size--and probably to a large portion of the difference that actually exists--between the jaws of savage and civilised man. If we consider, further, that concurrently with the diminished use of the jaws there was an increased use and development of the brain, it may well be that the process of reduction of the former was facilitated by the diversion of a portion of the supply of arterial blood to the latter.

These two causes--cessation of selection or *panmixia*, and the effect on the individual of greater or less exercise of the parts--are admittedly real causes, their effects can be roughly estimated, and they seem fully adequate to account for the comparatively small difference that actually exists between the jaws of the lowest and highest races now on the globe.

The question of purely individual variation, requires further consideration, because, owing to the unhesitating acceptance of the inheritance of acquired characters by Darwin and most contemporary naturalists, the important bearing of facts proving the effect of external conditions on individuals has been, and still is, altogether overlooked. Two cases in particular are continually quoted by the advocates of inheritance, as if they were in some way antagonistic to Weismann's theory, whereas they really support that theory, and almost prove it. The first is that of the Texan species of Saturnia (Emperor-moth) which feeds on the black walnut (*Juglans nigra*), and which, when pupæ were brought to Switzerland and the larvæ raised from the eggs laid by the moth were fed on the common walnut (*Juglans regia*), produced moths which differed so much from the parent both in colour and form that it appeared to be a new species. Darwin, writing to Wagner of this case, says: "When I wrote the *Origin*, and for some years afterwards, I could find little good evidence of the direct action of the environment; now there is a large body of evidence, and your case of the Saturnia is one of the most remarkable of which I have heard." In referring to this letter, Professor H. F. Osborn, in one of the most intelligent discussions of this question I have yet seen from an American author, says, "Darwin distinctly abandoned the utility principle in the case of Saturnia";[6] and Mr. D. G. Elliott, in his presidential address to the American Ornithologists' Union in 1891, quoted the same case as affording striking evidence of the transmission of acquired modifications.

Professor Lloyd Morgan (in his *Animal Life and Intelligence*, p 163-166) sees clearly that this and other cases do not prove more than a modification of the individual; but it seems to me to go further than this. For here we have a species, the larvæ of which for thousands, perhaps millions of generations, have fed upon one species of plant, and the perfect insect has a definite set of characters. But when the larvæ are fed on a distinct but allied species of plant, the resulting perfect insect differs both in colouration and form. We may conclude from this fact, that some portion of the characters of the species were dependent on the native food-plant, *Juglans nigra*, and that this portion changed under the influence of the new food plant. Yet the influence of the native food had been acting uninterruptedly for unknown ages. Why then had the resulting characters not become fixed and hereditary? The obvious conclusion is, that, being a change produced in the body only by the environment, it is not hereditary, no matter for how many generations the agent continues at work; in Weismann's phraseology, it is a somatic variation, not a germ variation.

The other case is that of the brackish water shrimp,

Artemia salina, which, by the water becoming gradually salter, was changed into what had before been considered a distinct species, *A. milhausenii*; and the reverse change was also effected, the modification in both cases being in proportion to the alteration in the salinity of the water, and therefore spread over two or three generations. Here too there seems to be no heredity, however long either form has been submitted to the influence of the modifying medium.

Similar to this is the case given by Mr. Thiselton Dyer (*Nature*, vol. xliii., 581) of the tissue-papery leaved *Arabis anachoretica*, which grows in hollows of rocks sheltered from sun and rain, but which, when cultivated at Kew, changed at once to the common *Arabis alpina*, of which it is a modification due to the action of the environment on the individual plant. How different is this from the behaviour of plants which have been developed by germ-variation and selection, such as most of the true alpine plants which retain their compact dwarf foliage and large flowers under such different conditions when raised from seed in our gardens! Other species, however, which have a considerable range and become dwarfed in some localities by adverse conditions of drought, wind, or other causes, at once grow to a larger size when cultivated and sheltered. Many of our British botanists are now applying this test to distinguish those forms of our native plants which owe their peculiarities to germ-variation from those which have been modified individually by the action of the environment. We are often unable to decide by mere observation to which class any particular variety of local form belongs, but cultivation at once determines the point; for while the former transmit their peculiarities, however minute these may be, to their offspring, the latter revert at once to the parent form. Mr. Beeby has proposed to call the former class "intrinsic," the latter "extrinsic" varieties, useful terms which indicate that the one are due to an internal cause, are therefore stable, and show us the incipient stage of species-formation; while the other is merely an external modification of the individual which has no stability, being wholly due to the direct action of the environment. It is very important to note the sharp distinction between these two kinds of varieties, externally so alike though having a fundamentally different origin. There appears to be no gradation from one to the other. The individually acquired or *extrinsic* character, however long it may have persisted, disappears instantly when the special environment that produced it is changed, as in the case of the Texan Saturnia, the papery-leaved Arabis, and many similar cases; while *intrinsic* characters--those due to germ-variation--however slight they may be, as in the various races of mankind, many of the closely allied species of moths, and some of the sub-species or varieties of our native plants, preserve their characteristic features under greatly changed conditions.

The cases now given of change in the individual due to

external causes and often of a very marked character, render it exceedingly probable that a large portion of the observable difference in the size of the jaws of civilised man and of some domesticated animals, as well as all those changes produced more or less suddenly by a change in the environment, are mere individual effects which are not hereditary; while, whenever such changes appear in species that have long been subjected to uniform conditions--as in the case of the Texan Saturnia--they indicate that some portion of the external peculiarities of those species were individually acquired characters, and therefore afford strong evidence of the non-inheritance of such characters.

I have now fairly met, so far as the space at my disposal will allow, the strongest arguments of the advocates of use-inheritance as a law of nature and as a factor in evolution. I have shown that the effects which it ought to produce in the case of mankind do not appear, and that breeders of animals do not recognise it as a factor to be taken account of. The alleged cases of inherited habits or instinct, supposed to be useless, are shown in one case to be not necessarily so; while all such cases involve so many elements of uncertainty or ignorance that no conclusion of value can be drawn from them. The alleged difficulty of the origin of horns except by the inherited effect of blows and pressures, I have shown to be founded on error as to fact; and their origin by normal variation, and development where useful by selection, to be supported by the frequent occurrence of dermal excrescences in many animals. The case of the mammalian teeth has been shown to be quite explicable without use-inheritance, the mode of action of which is, in this case, itself inexplicable. Mr. Herbert Spencer's three main arguments to prove the inadequacy of natural selection have been fully discussed, and have, I venture to think, been shown to be entirely inconclusive, since they are either founded on comparatively unimportant and adventitious facts, or on a neglect of some of the most important conditions under which natural selection in its various forms comes into play.

In order to render the position clearer I wish, before concluding, to say a few words on the general question. Those who are termed Neo-Darwinians do not yet maintain that use-inheritance (to use Mr. Ball's convenient term) does not exist, but merely that it has not been proved to exist. Whether it actually occurs--and if it occurs at all, I believe, it must occur constantly--can only be proved either by very careful and long-continued experiment, or by the demonstration of some theory of heredity which either necessarily includes or excludes it. But even if it does exist I myself believe that it is altogether unimportant as a factor of evolution, and that we have evidence sufficient to prove that natural selection is not inadequate for want of it.

When we urge that the effects of use-inheritance, if it exists, ought to be abundantly visible in some such ways as I have suggested in the early portion of this article, its advocates reply, that only a small portion of what is acquired by the parent is transmitted to the offspring, and that its effects may, therefore, only become visible after a long series of generations; but as it is necessarily cumulative it must produce a considerable result in the course of ages. All we can say in reply to this is, that it is pure hypothesis, and that, if true, it may serve to explain the difficulty of obtaining evidence in its favour, but at the same time it indicates that use-inheritance can be of no value as a factor in evolution. Variation is so large and so constant that any required character can be greatly modified in a very short time, of which Darwin gives many illustrations. The comb of the Spanish cock was made upright, the comb and wattles of the Polish fowl were completely abolished, and the average weight of ducks was raised from four pounds to six pounds in a few years. Our sheep, pigs, and cattle were wonderfully improved, and often completely changed in form between the latter portion of the eighteenth and the first half of the present century, while many of our flowers and fruits have been nearly doubled in size, and greatly improved in form and colour in the same period. But natural selection has a great advantage over artificial selection in the enormous scale on which it works, giving much greater scope for the occurrence of favourable variations in large numbers. Thus, whenever some great change of conditions led to a more severe struggle for existence, the modifications of structure needed for adaptation to the new environment would soon be effected without any aid from use-inheritance.

That this is the fact is further indicated by the large range of characters and adaptations that must have been produced by variation and selection alone, since use-inheritance cannot possibly have had any part in their development. Mr. Spencer admits that there are many such, but does not recognise the weight of the argument which they afford against the need of use-inheritance as a factor in evolution. It is well, therefore, briefly to enumerate some of the more important of them. We have already referred to the teeth, in their numerous peculiarities of form, structure, and mode of growth, many of which are quite removed from any direct influence of use in their production. Still less can we impute the hair to such causes, in its varieties of length, thickness, texture, and colour, with its occasional modification into protective armour, such as plates, scales, or spines, or into offensive weapons, such as the horns of the rhinoceros. The bill of birds, in such strange modifications as are presented in the duck, the spoonbill, the heron's spear, the woodpecker's chisel, the snipe's sensitive borer, the enormous but very light bill of the toucan, the powerful nut-cracking bill of the cockatoo, and many others, all evidently adapted to special uses, but by no possibility developed by those uses; the wonderful modifications of the stomach in ruminants, and especially in the camels; the whole series of protective, warning, and recognition colours of animals; the numerous peculiarities of structure and instinct in neuter insects where use-inheritance is absolutely excluded; and, lastly, the whole of the wonderful protective and distributive contrivances of fruits and seeds, and the still more wonderful and more complex adaptations of flowers to insect fertilisation, this latter, be it remembered, not under the pressure of an individual struggle for life, but only apparently to obtain an increase of vigour and a somewhat more rapid multiplication, giving an advantage over other species in the general struggle for existence.

Even more removed from possible development through use are the special organs of sense, such as the eye and the ear. The complexity of structure in the internal ear is amazing, with its spiral and semi-circular canals and tubes, its membranes, ducts, and cartilages, its extraordinary stirrup, hammer, and anvil bones, its wonderful rods, hair-cells, and otoliths. It seems like some strange machine, the connection of whose various parts and their mode of action it is impossible to follow. Yet we feel sure that every detail has its use; and we have here an organ the co-adaptation of whose several constituent parts is essential to its utility, and is apparently more difficult to bring about by variation and selection than in the cases where Mr. Spencer thinks it absolutely necessary to call in the aid of use-inheritance. But in this case it is utterly unimaginable that any amount of air-waves impinging on the tympanum can have tended directly to the production of this highly complex and delicately adjusted organ. The case of the ear alone appears to me sufficient to prove that use-inheritance is not required for the development and progressive modification of the most complex and beautifully adjusted structures.

We thus see that there is a wide range of characters and structures, often involving the most beautiful adaptations, in which use-inheritance can admittedly have no share; that even if it exists in other cases it is unnecessary, since it can only give a little help in a process which is demonstrably within the power of variation and natural selection; and that the strongest arguments that have been urged, either to show its supposed necessity or to prove its actual existence, break down on close examination, and in some cases even afford strong evidence against it. Our conclusion, therefore, is, that no case has yet been made out for the inheritance of individually acquired characters, and that variation and natural selection are fully adequate to account for those various modifications of organisms which have been supposed to be beyond their power.

Notes Appearing in the Original Work

[1] I discussed this aspect of the question in an article on "Human Selection," *Fortnightly Review*, Sept.,

1890; and, further, in the *Arena* of Jan., 1892, under the title of "Human Progress: Past and Future." on
[2] *The Germ Plasm: a Theory of Heredity*, by August Weismann. Walter Scott. London, 1893. on
[3] *Evolution of the Colours of North-American Land Birds*, by Charles A. Keeler, California Academy of Sciences, San Francisco, January, 1893. on
[4] I have omitted Professor Ray Lankester's suggestion, of a process of selection owing to those individuals with more perfect eyes occasionally finding their way out, because the ancestral immigrants were probably carried far into the caverns by torrential floods, and could only escape by following the water to some of its outlets, success in which would not depend on special acuteness of vision. on
[5] *Origin of Species*, 177, and more fully and with admirable force and clearness in *Animals and Plants under Domestication*, vol. ii. 221. on
[6] *Are Acquired Variations Inherited?* An Argument, by Henry Fairfield Osborn, in *The American Naturalist*, February, 1891. on

www.ingramcontent.com/pod-product-compliance
Lightning Source LLC
Chambersburg PA
CBHW022054190326
41520CB00008B/784